KETOCONAZOLE

The Best Guide for Effective Treatment of Fungal Infections Such as Eczema, Joke Itch, Athlete's Foot, Vaginal Candidiasis, Oral Thrush, Skin Rash, Ringworm, etc, and be Side Effect-Free

Juana Porro

Table of Contents

Introduction

In the medical world, ketoconazole is an important medicine known as an antifungal drug used to treat a variety of health issues. It is used in the treatment of a wide range of fungal infections, including chronic skin infections such as athlete's foot, toenail fungus, jock itch, and ringworm. It is also used to treat certain types of yeast and other fungal infections affecting the throat and lungs, such as thrush and HACEK.

Ketoconazole is used for other conditions as prescribed by your doctor. This article will discuss the various uses of ketoconazole, the side effects, storage requirements, warnings & precautions, instructions for taking ketoconazole, as well as its potential drug interactions and overdose & missed doses.

What is Ketoconazole?

Ketoconazole is an antifungal medicine used to treat a variety of infections caused by fungi like athlete's foot, toenail fungus, ringworm, etc. It is also used to treat fungal infections of the throat and lungs such as thrush and HACEK and other conditions as prescribed by your doctor. Ketoconazole works by blocking the growth of fungi. It is classed as an azole antifungal, which works by stopping the fungi from reproducing by blocking the synthesis of an important component called cell membrane sterols.

Uses of Ketoconazole

Ketoconazole is used to treat the same range of fungal infections as the antifungal medicine, but is often prescribed for more serious infection cases that don't respond to the regular antifungal. Other than the aforementioned skin and chest infections, it is used to treat serious fungal infections affecting the mouth, throat, esophagus, stomach, intestine, bone, urinary tract, blood, and reproductive organs. It may also be used as part of the preparation for organ transplantation.

Side Effects of Ketoconazole

Although ketoconazole is usually well tolerated, there are some side effects that could occur during the course of treatment. These include nausea, vomiting, diarrhea, gas, stomach upset, headache, dizziness, and abdominal pain. More serious side effects associated with ketoconazole include liver problems, liver inflammation, hepatitis, blood in urine, yellowing of the eyes or skin, hallucinations, confusion, fever, rash, and itching.

Instructions for Taking Ketoconazole

Ketoconazole should be taken as prescribed by a doctor and the instructions for use may differ from patient to patient. Generally, it is taken orally, either in pill form or as a liquid suspension. The dosage will vary depending on the particular infection being treated and the patient's age and weight. It is important to follow the instructions and not exceed the prescribed dosage.

Warnings & Precautions

Before taking ketoconazole, medical history should be provided to the doctor along with any allergies to the drug and/or any other medications. It is important to inform the doctor of any medical conditions or if the patient is pregnant or breastfeeding. Also, patients with liver disease or risk factors for liver disease, such as heavy drinking, are not advised to take ketoconazole. Other drugs and supplements should be discussed with the doctor, as they may interact with ketoconazole.

Potential Drug Interactions

Ketoconazole can interact with other medications and supplements, and it is important to inform the doctor of any other drugs that the patient is taking. Other antifungal drugs can interact with ketoconazole, and it is important to not use them together. Ketoconazole may also interact with antidepressants such as monoamine oxidase inhibitors (MAOIs), tricyclic antidepressants, selective serotonin reuptake inhibitors (SSRIs), and other supplements such as aspirin, birth control pills, and the herb St. John's wort.

Overdose & Missed Doses

If an overdose of ketoconazole is suspected, the doctor should be contacted immediately. To avoid overdosing, it is important to take the medication as prescribed by the doctor. If a dose is missed, it is important to take it as soon as possible. However, it is important to note that taking a double dose to make up for a missed dose can be dangerous.

Storage Requirements

It is important to keep ketoconazole at a temperature below 30°C and away from moisture and heat. The medicine should also be stored in a childproof container and kept away from children.

Benefits of Ketoconazole

Ketoconazole is a useful medication that can help treat various fungal infections. It is also effective in preventing fungal infections from reoccurring. The drug is relatively easy to take, and is usually well tolerated by most people. It also has fewer long-term side effects than some other antifungal medications.

Ketoconazole is an antifungal medication that is most commonly used to treat fungal infections. It is commonly used to treat ringworm, jock itch, athlete's foot, dandruff, and other fungal infections. As studies continue to show the effectiveness of this antifungal, more people are beginning to understand the many benefits of ketoconazole.

One of the main benefits of ketoconazole is that it is effective in treating a wide variety of fungal infections. This includes athlete's foot, ringworm, jock itch, and dandruff. It has been tested to be highly effective in treating these fungal infections. In addition, ketoconazole has been shown to be effective in treating certain types of yeast infections, which are also caused by fungus. This drug is also used to treat certain types of skin infections and fungal nail infections, such as onychomycosis. Additionally, ketoconazole is often used to treat fungal infections that affect internal organs, such as candidiasis.

Another benefit of ketoconazole is that it is well tolerated and usually does not cause many side effects. Ketoconazole is generally considered to be a safe drug with few side effects. Most people who take it do not experience many if any side effects, which is why it is so popular among many people.

Ketoconazole is also known to be effective at fighting fungal infections that are resistant to other antifungal medications. This is because the drug targets multiple parts of the fungal cell, making it harder for them to develop resistance to it. This makes it a reliable medication for treating fungal infections that are highly resistant to other antifungal drugs.

Other benefits of ketoconazole include it being relatively easy to use and it is available as a cream, or it can be taken orally. The cream is applied directly to the affected area twice a day until the infection clears up. Oral ketoconazole is taken one to four times a day, depending on the severity of the infection. This makes it a convenient medication to take, especially for those who suffer from chronic fungal infections.

One of the most important benefits of ketoconazole is that it can help prevent new fungal infections from forming. This is because ketoconazole helps to keep the number of fungal cells low, which prevents them from multiplying and causing further infections. This makes it a great option for people who want to reduce their risk of developing fungal infections or who already suffer from frequent fungal infections.

In addition, ketoconazole is not known to interact with other medications, making it an ideal choice for people who are taking multiple medications or who have an interaction between certain medications. It is also safe to use during pregnancy and breastfeeding, making it a great option for expectant mothers who need to find an effective medication to treat their fungal infection.

Overall, ketoconazole is an effective and safe antifungal medication that can be used to treat a variety of fungal infections. It is well-tolerated, easy to use, and has few side effects, which makes it a popular choice for many people. In addition, ketoconazole is known to be effective at treating resistant fungal infections and also helps to prevent new fungal infections from forming.

Conclusion

Ketoconazole is a helpful antifungal medication used to treat various fungal infections. It can be taken orally, in either pill or suspension form, and is usually safe and well tolerated by most people. Certain conditions and potential drug interactions should be discussed with the doctor before starting treatment with ketoconazole. In order to prevent overdose or missed doses, it is important to follow the doctor's instructions and keep the medication away from children. In conclusion, ketoconazole is a useful medication in helping to treat various fungal infections.

THE END

Made in the USA
Las Vegas, NV
20 April 2024

88936777R00015